AF297597

SUR LES PROPRIÉTÉS RÉFRACTAIRES DE LA SILICE

SUR LES PROPRIÉTÉS RÉFRACTAIRES DE LA MAGNÉSIE

BY

MM. H. LE CHATELIER ET B. BOGITCH.

From the Transactions of The Ceramic Society, 1917—18.
Vol. XVII.

I.—Sur les propriétés réfractaires de la Silice.

par MM. H. Le Chatelier et B. Bogitch.

L'EMPLOI des briques de silice dans la construction des fours a seul permis la généralisation du chauffage à chaleur régénérée de W. Siemens. Avec les briques d'argile, les voûtes ne pouvaient supporter les températures ainsi obtenues et s'effondraient rapidement. Jusqu'ici cependant, on n'a pas expliqué d'une façon certaine cette supériorité de la silice sur l'argile.

Nos expériences récentes sur les propriétés réfractaires de l'argile[1] permettent au contraire d'entrevoir une explication très précise. Les briques d'argile, comme nous l'avons établi, commencent à se ramollir entre 1,300° et 1,400°. Elles ne peuvent plus alors supporter d'efforts importants sans céder d'une façon continue et indéfinie Elles le font d'autant plus rapidement que la température est plus élevée, c'est à dire, elles se comportent comme une matière vitreuse, elles n'ont pas de véritable point de fusion, mais seulement un palier très étendu de fusibilité. Le prétendu point de fusion de la kaolinite pure, fixé à 1,780° et identique à celui du quartz, déterminé de la même façon par l'emploi de montres de Seger, correspond en réalité à l'affaissement rapide de la matière sous une charge égale au poids même de l'éprouvette, c'est à dire de l'ordre de 1 gramme par cm². Sous une charge 10,000 fois plus forte, c'est à dire de 10 kilogs. par cm² l'affaissement du kaolin se produit avec la même vitesse, 400° plus bas.

Pour expliquer la façon différente de se comporter de l'argile et du quartz, on pouvait supposer que ce dernier présentait au contraire un véritable point de fusion, sans ramollissement préalable, indépendant par suite de la pression. Nos expériences ont pleinement confirmé cette prévision. Quand on écrase vers 1,500° un petit cylindre d'argile, on le voit gonfler en forme de tonneau, puis s'aplatir en une galette à bords arrondis, sans présenter à aucun moment de rupture proprement dite. Après refroidissement, la masse écrasée a conservé toute sa dureté première. Au contraire, avec la silice,

---

[1] C. R., CLXIV, 904, 1917.

A

la première action de la pression ne produit aucun effet appréciable. Mais en l'augmentant progressivement, on voit brusquement l'éprouvette se briser, en présentant les deux cônes de glissement que l'on observe dans la rupture par compression de toutes les matières dures. Les fragments brisés ne se ressoudent aucunement pendant le refroidissement. L'effort nécessaire pour produire cette rupture brusque décroît progressivement avec l'élévation de la température.

Le tableau et la courbe (Fig. 1) ci contre résument les résultats de nos expériences sur une bonne brique de silice de fabrication Américaine, marque "Star." Les résistances à l'écrasement sont exprimés en kilogs par cm².

Température	Résistance
15°	170
520°	158
670°	150
800°	139
950°	125
1050°	120
1200°	85
1320°	62
1460°	50
1540°	37
1600°	30

Ces nombres conduisent par extrapolation à une résistance de 12 kilogs à 1,700°, température habituelle des voûtes de fours d'aciéries. Cette résistance correspond à peu près à dix fois l'effort que les briques supportent dans les voûtes. La stabilité de ces dernières est donc assurée.

Cette résistance mécanique se conservant jusqu'à des températures élevées est une particularité toute spéciale aux briques de silice. On ne la retrouve pas, non seulement dans les briques d'argile, mais même dans les briques de magnésie, dont le constituant essentiel, la magnésie, est cependant bien moins fusible encore que la silice.

Voici la raison de ces différences. Dans les deux cas, les briques renferment des oxydes basiques étrangers à la matière réfractaire principale. Ces oxydes : Alumine, chaux, oxyde de fer, alcalis, etc., donnent dans tous les cas naissance à une matière fusible, déja liquide aux environs de 1,200°. Dans le cas de la magnésie, les grains solides de ce corps nagent dans le magma fondu et glissent facilement les uns sur les autres comme le ferait du sable mouillé. La silice au contraire, du moins dans les briques bien cuites, forme un réseau continu dans les pores duquel se loge la masse fondue, comme l'eau se loge dans les pores de la pierre ponce, sans en diminuer la résistance mécanique. La formation de ce réseau, conséquence de la recristallisation de la silice, est due, comme l'un de nous l'a

fait voir,[1] aux différences de solubilité des différentes varié
tés allotropiques de la silice. Le quartz, instable à haute
température, se dissout dans le magma fondu et recristallise
à l'état de cristobalite d'abord, puis de tridymite (Fig. 2). Cette
recristallisation de la silice et par suite le formation du réseau
indéformable, exigent une cuisson effectuée à une température
convenable et suffisamment prolongée. Les briques peu cuites,
dont le réseau n'est pas encore formé, sont composées de grains
de quartz nageant dans la masse fondue; elles sont plastiques,
comme les briques de magnésie ou d'argile. C'est là un fait
bien connu dans les usines; les briques peu cuites sont fusibles
et inutilisables.

Lorsque l'on chauffe une bonne brique de silice, sa
résistance diminue cependant avec l'élévation de la température.
Cela tien à ce que la solubilité de la silice croissant avec la
température, il se produit une dissolution progressive du réseau
cristallin tendant à le désagréger et même, à le disloquer
complètement. Cet effet sera d'autant plus retardé que le
réseau sera mieux développé. C'est de là que dépend avant
tout la qualite des briques de silice. L'étude des facteurs dont
dépend la rigidité de ce réseau présente donc une importance
de premier ordre. Les recherches dont nous rendons compte
aujourd'hui ont pour objet d'éclaircir ce problème.

Les principaux facteurs à envisager sont:
La proportion des fondants.
La température actuelle de la brique.
La bonne formation du réseau.
La désagrégation du réseau par gonflement ultérieur.

Proportion des oxydes fondants. Nous avons analysé
quelques briques qui nous avaient été signalées comme ayant
donné des résultats particulièrement satisfaisants à l'emploi. Le
tableau ci dessous donne le poids p. cent. des oxydes basiques,
ainsi que le poids total des sulfates obtenus par une attaque
à l'acide fluorhydrique, suivie d'une évaporation à sec après
addition d'acide sulfurique:

Marque	Al_2O_3	Fe_2O_3	CaO	MgO et K_2O	Total	Sulfates
Star	0·94	0·15	1·79	0·50	3·38	9·06
Assailly	2·72	1·30	0·20	0·25	4·47	13·60
G. I.	1·02	0·60	1·48	0·00	3·10	8·40

[1] H. Le Chatelier, *Rev. de Mettallurgie, XIII.*, 330, 1917.

La brique Star est celle dont la résistance a été donnée plus haut.   La brique d'Assailly, dont la fabrication remonte à 30 ans, avait séjourné un an dans la paroi d'unconduit de gaz d'un four d'aciérie. C'est sur cet échantillon que l'un de nous(¹) ·a reconnu pour la première fois la transformation du quartz en tridymite dans les briques longtemps chauffées. Enfin la brique G.I. est une brique de fabrication française, passant pour une des meilleures de celles que nous produisons actuellement.

D'après ces chiffres les bonnes briques de silice renferment entre 3 et 5 p. cent. d'oxydes basiques et le poids des sulfates varie de 8 à 14 p. cent. Le rapport du poids des oxydes à celui des sulfates varie nécessairement suivant la nature des bases.   Mais comme leurs proportions relatives restent généralement comprises entre des limites assez reserrées, on peut admettre que le poids des oxydes représente une fraction sensiblement constante du poids des sulfates, soit en moyenne 35 p. cent.   Ce dosage des sulfates peut être fait rapidement et suffit pour apprécier la composition d'une brique de silice.

Température.   La température que supporte la brique dépend entièrement de l'usage auquel elle est destinée. Dans les fours à acier, elle doit supporter une température de 1,700°. Les briques présentant la composition cidessus indiquée, possèdent, si elles ont été bien fabriquées, une résistance suffisante à la température en question. On emploie encore les briques de silice dans les fours à distiller la houille, où la température est moins élevée. On peut sans inconvénient accepter pour cet usage une proportion au moins double des oxydes basiques, ce qui facilite grandement la fabrication.

Constitution du réseau.   C'est là la partie la plus délicate de la fabrication.   Pour développer le réseau, il faut maintenir très longtemps la brique à une température où le magma fondu soit suffisamment fluide.   L'expérience semble indiquer, comme conditions les plus favorables, un chauffage prolongé plusieurs jours à une ·température voisine de 1,450°.   Cette température doit être inférieure à celle à laquelle le quartz employé se transforme directement et rapidement en cristobalite. Le réseau se forme exclusivement aux dépens des parties de silice qui recristallisent par dissolution passagère dans la masse fondue. Enfin cette recristallisation sera d'autant plus complète et plus rapide, toute chose égale d'ailleurs, qu'il y aura dans le mélange soumis à la cuisson plus de quartz fin et même très fin.   Il faut cependant une certaine proportion de gros grains pour empê cher la formation de fentes dont la propagation se fait trop facilement dans les matière uniformément fines.

---

(1) H. Le Chatelier, C.R., CXI, 123, 1890.

Désagrégation du réseau. Lorsqu'il reste après cuisson des grains de quartz non transformés et que l'on chauffe ensuite brusquement la brique à une température à laquelle le quartz se transforme rapidement, le gonflement qui accompagne cette transformation brise le réseau et enlève toute solidité à la brique. De plus, la pression qu'elle supporte dans la voûte s'oppose à son gonflement latéral et produit le phénomène *d'écaillage*; elle tombe alors par petits morceaux, entraînant parfois en quelques jours la destruction d'une voûte qui aurait dû faire des mois de service. La photographie ci contre (Fig. 3) montre une brique semblable avant et après mise en place dans le four. On voit très nettement les fentes provoquées par ce gonflement. Ce défaut est peut être le plus grave et certainement le plus fréquent de ceux que présentent les mauvaises briques de silice. Dans la fabrication normale d'une brique bien cuite, le même gonflement se produit au moment de la transformation directe des gros grains de quartz qui ne se dissolvent jamais que sur une très faible épaisseur, mais l'inconvénient n'est pas le même, par ce qu'alors le gonflement de la brique peut se développer librement dans tous les sens et que d'autre part le phénomène est assez lent pour permettre au réseau de se reformer sur les points ou il a pu être brisé. Néammoins ce phènomène inévitable diminue notablement la résistance mécanique de la brique.

Voici maintenant le détail des expériences qui ont conduit aux conclusions précédemment énoncées. Elles ont porté sur des briques qui nous ont été fournies par les usines de la Marine, fonderie de cannons de Ruelle et fabrique de blindages de Guérigny, et pour lesquelles on nous a donné l'indication des qualités à l'emploi. Nous avons mesuré la proportion des sulfates, les densités absolues et apparentes la résistance à l'écrasement en kilogs par cm² à des températures déterminées et après un temps de chauffage également donné.

Marques	Qualité	Sulfates	D. Absolue	D. Apparente	Résistance		
					Température	Temps.	R
Assailly	T. Bonne	13·6%	2·30	1·92	15°		550
					1600°	60'	90
Star ..	T. Bonne	9·06%	2·33	1·66	15°		170
					1600°	5'	33
					Id	60'	30
G.I. ..	T. Bonne	8 40%	2·33	1·88	15°		185
					1600'	60'	41
R.B. ..	T. Bonne	13·1%	2 35	1·60	15°		62
					1600°	60'	9·5
R.L. ..	T. Bonne	14·3%	2·40	1·85	15°		265
					1600°	5'	41
					Id	60'	25
G.A. ..	Bonne	14·0%	2·40	1·77	15°		190
					1600°	60'	21
D. ..	Bonne	8·4%	2·45	1·73	15°		320
					1600°	5'	55
					Id	60'	20
G.A.I.	T. Mauvaise	14·5%	2·46	1·80	15°		252
					1600°	60'	4·4
R.F. ..	T. Bonne	13·7%	2·48	1·84	15°		195
					1600°		11
G.A.2 ..	T. Mauvaise	12·8%	2 48	1·78	15°		148
					1600°	60'	5
G.M. ..	Médiocre	9·5%	2·53	1·84	15°		84
					1600°	5'	17
					Id	60'	2
R.L. ..	Mauvaise (éclate)	9·75%	2·56	1·94	15°		350
					1600°	5'	18
					Id	60'	4·5
R.S.G.	Mauvaise (fond)	25·0%	2·56	1·73	15°		57
					1550°	60"	22
					1600°		fondue

Ces résultats permettent de formuler une conclusion pratique très précise. Toutes les bonnes briques ont à 1,600°, après une heure de chauffage, une résistance à l'écrasement au moins égale à 10 kil. par cm². La prolongation du chauffage à cette température diminue peu leur résistance, à l'inverse de ce qui se produit pour les mauvaises briques.

DISCUSSION.

M. Henry LE CHATELIER est très heureux d'avoir provoqué par sa communication une discussion aussi complète ;

il est sur presque tous les points d'accord avec les auteurs des observations faites.

Peut-être n'est-il pas nécessaire de poursuivre de nouvelles recherches sur la densité des diverses variétés de silice, aux températures élevées, comme le demande le Dr. Scott. On peut, en effet, les calculer très exactement en partant des densités mesurées à la température ordinaire et utilisant les coefficients de dilatation. On trouve ainsi que la cristobalite possède vers 1,000°, une densité voisine de 2·22. Il reconnait au contraire la difficulté de déterminer quantitativement les proportions de tridymite et de cristobalite et, plus encore l'impossibilité d'obtenir la transformation complète du quartz en tridymite, dans la première cuisson des briques. Cette transformation demande des semaines pour s'achever dans le four à acier, où la température est cependant bien plus élevée.

Le Professeur Fearnsides signale un fait tout à fait exact au sujet de la transformation première du quartz en cristobalite sous l'action de la chaleur. Après les échauffements de peu de durée, suivis d'un refroidissement rapide, on n'observe jamais que de la cristobalite. Ainsi, dans la paroi d'une cornue Bessemer, on trouve de beaux cristaux de cristobalite; mais pas ou peu de tridymite, parce que les chauffages sont peu prolongés et les refroidissements rapides. Ces conditions sont toutes différentes de celles du four Siemens, où la température est maintenue presque constamment très élevée. C'est alors seulement que la tridymite se développe régulièrement.

A la question posée par Mr. G. W. Mottram, au sujet de la calcédoine (flint), on peut répondre que cette variété de silice se transforme comme le quartz, d'abord en cristobalite, mais elle le fait à une température très basse, à partir de 1,150°, bien avant le commencement de la fusion de silicates. En l'absence de liant, le gonflement devient énorme et la brique très poreuse perd toute résistance. Il ne semble pas que jusqu'ici on ait réussi à fabriquer de bonnes briques de silice en employant exclusivement de la calcédoine.

Il n'est évidemment pas impossible d'employer directement des briques en silico-calcaire à faible teneur en chaux, à condition de conduire très doucement l'échauffement du four, de façon à faire sur place, la cuisson de la brique. Mais il est douteux que cette pratique soit avantageuse, parce que, si les parties les plus chauffées cuisent bien, les portions intermédiaires de la brique, celles qui supportent des températures comprises entre 500° et 1,200°, perdent toute solidité et sont alors sujettes à se désagréger sous l'action des mouvements inévitables dûs à la dilatation des parois chauffées.

La question de savoir si l'on peut remplacer les grains anguleux d'une roche de quartz broyé, per du sable en grains ronds, demande encore de nouvelles études. Des essais en cours, faits avec un mélange de 75 per cent. de sable de Fontainebleau et 25 per cent. de silice impalpable, passée au tube broyeur, ont donné des résultats encourgeants.

Mr. J. Holland proteste avec beaucoup de raison contre l'idée de chercher à fabriquer des briques de silice entièrement transformées en tridymite. L'auteur du mémoire n'a jamais soutenu une semblable opinion. Cela serait absolument irréalisable, puisqu'il faut pour cette transformation un séjour de 1 mois dans un four d'aciérie à une température supérieure à 1,600°, quand les plus fortes cuissons industrielles des briques de silice, n'atteignent pas 1,400°, et que la cuisson à la température la plus élevée ne dure souvent que quelpues heures, jamais plus de quelpues jours. Il semble nécessaire d'obtenir cependant une proportion suffisante de tridymite, ce à quoi on peut arriver par l'introduction dans la pâte d'une proportion suffisante de quartz impalpable, c'est-à-dire traversant le tamis de 4·900 mailles au cm².

L'indication donnée par Mr. Cecil Desch au sujet de la fusibilité de la silice demanderait à être discutée de très près. Les diverses variétés de silice: quartz, cristobalite et tridymite ont nécessairement des points de fusion différents, la tridymite fondant à la température la plus élevée. Il semble difficile d'admettre que son point de fusion puisse s'abaisser jusqu'à 1,625°. On ne l'observe même pas à 1,700°. On voit bien dans les vieilles briques complètement transformées en tridymite et chauffées à cette température, une partie de la masse devenir vitreuse par dissolution d'une certaine quantité de tridymite dans les silicates métalliques; mais les grains de tridymite, dans les parties respectées par le fondant restent absolument inaltérées comme structure et dimensions des cristaux. On cite parfois aussi pour le point de fusion de la variété quartz, un point de fusion de 1,480°. Ce chiffre ne parait pas davantage établi sur des bases précises. Il est probable que le point de fusion de la tridymite, ou si l'on préfère, son passage à l'état amorphe est voisin de 1,800°.

Mr. A. Lindsay Forster a décrit avec beaucoup de précision les apparences successives d'une même brique de silice examinée dans des régions de plus en plus éloignées de la portion chauffée. La voûte des fours d'aciérie est généralement impregnée, sur la partie inférieure, de scories ferrugineuses et c'est dans cette région que les cristaux de tridymite sont le plus complètement développés.

Mr. W. Donald est effrayé du poids des données scientifiques sous lesquelles on veut écraser les fabricants de briques de silice. Sans contester aucunement les résultats auxquels sont arrivés ces fabricants par une longue expérience, s'étendant sur plusieurs générations, il sera permis de faire remarquer que, en l'absence de cette expérience pratique, les cherches de laboratoires arrivent, par leurs études scientifiques, à produire après quelques mois de travail, d'aussi bonnes briques que celles dont la mise au point, par les méthodes purement empiriques, a demandé de longues années de tàtonnements. C'est bien là, le but essentiel des méthodes scientifiques de travail. Quant aux différences signalées entre les briques d'Ecosse et de Sheffield, il serait sans doute facile d'en reconnaitre rapidement la cause, si l'on avait en mains les échantillons à comparer. Il n'est nullement prouvé que, pour avoir de bonnes briques, il faille employer des roches d'une nature spéciale. Les études faites jusqu'ici montrent, au contraire, qu'avec tous les quartz d'une dureté suffisante, et ne se pulvérisant pas au feu, il est possible, et même facile, de faire de bonnes briques à condition d'employer une cuisson approprié à chaque nature de pierre.

Les explication de Mr. Mellor montrent très nettement quel est le but des études scientifiques et comment on peut les traduire pour les rendre accessibles aux fabricants. Peut-àtre, cependant, n'a-t-il pas assez insisté sur une particularité de grande importance. L'une des variétés de silice, à faible densité, présente une propriété très curieuse ; mais aussi très nuisible. Elle éprouve vers 225° un changement brusque de dimensions qui amène des ruptures, quand on chauffe trop rapidement les briques. On ne saurait assez recommander aux métallurgistes de mettre très lentement en feu les fours construits en briques de silice. Faute de cette précaution, on s'expose à ruiner, dès les premiers moments du chauffage, des fours construits avec des briques excellentes à tous les autres points de vue.

En ce qui concerne l'affirmation de Mr. Fenner, que la tridymite cesse d'ètre stable aux températures les plus élevées des fours, on peut dire que la pratique journalière des fours d'aciérie l'a depuis longtemps contredite. Les parties les plus chauffées des briques, primitivement composées d'un mélange de quartz et de cristobalite avec trés peu de tridymite, se transforment progressivement et d'une façon complète en cette dernière variété de silice. L'expérience est d'ailleurs très simple à réaliser au laboratoire. Il suffit de prendre une brique presque complètement à l'état de cristobalite et de la chauffer pendant quelques heures à 1,700°, pour provoquer un

développement abondant de tridymite, et cela, sans aucune des
alternatives de chauffage et de refroidissement auxquelles Mr.
Fenner fait allusion. Ses expériences personnelles n'ont jamais
donné la preuve du contraire. Elles sont certainement très
exactes, mais ont été mal interprètées par leur auteur. En
chauffant, à une température élevée, de la tridymite, en présence
d'un fondant, et refroidissant lentement la masse, on y trouve
certainement de la cristobalite, même de la cristobalite très
bien cristallisée, comme j'en ai donné de nombreux exemples.
Mais cette tridymite ne s'est pas produite à chaud ; elle a
cristallisé pendant le refroidissement, en se séparant du magma
fondu dans lequel un excès de silice s'est dissout, grâce à
l'élévation de la température. Toutes les fois que l'on refroidit
avec une vitesse moyenne, un merre très chargé de silice, la
première variété qui cristallise est la cristobalite. De même,
si l'on chauffe de l'iodure rouge de mercure dans l'alcool
bouillant, c'est à dire bien audessous du point de transformation
de l'iodure dans sa variété jaune, et que l'on refroidisse ensuite
la solution, on produit toujours la variété jaune.   On n'en
conclut pas que le point de transformation, de l'iodure rouge
en iodure jaune, est audessous du point d'ébullition de l'alcool.
Il correspond, en réalité, à la température de 125°.   Il en
exactement de même avec la cristobalite.   Pour   rendre
l'expérience de Mr. Fenner concluante, il eut fallu tremper
l'échantillon, après le chauffage, pour empècher toute cristal-
lisation nouvelle de silice, au refroidissement.

[TRANSLATION.]

THE use of silica bricks, in the construction of furnaces,
has alone permitted the generalisation of firing with the
regeneration of heat. With clay bricks the arches could
not bear the temperatures thus obtained and would give way
in a very short time. Up to now, however, the superiority
of silica over clay has not been explained in any certain manner.

Our recent experiments, concerning the refractory properties of clay([1]) allow us, on the contrary, to perceive a very
precise explanation. The clay bricks, as we have proved, begin
to soften between 1,300° and 1,400°. They cannot, henceforth,

---

(1) *Compt. Rend.*, **184**, 904, 1917.

sustain important efforts without giving way in a continuous and indefinite manner. And this occurs more rapidly as the temperature is higher, that is to say, they act like a vitreous substance; they have no real fusing point but only an extensive fusibility range. The assumed point of fusion of pure kaolinite, fixed at 1,780° and identical with that of quartz, determined in the same way by the use of Seger cones, corresponds in reality to the rapid giving way of the substance under a load equal to the weight of the test piece, that is to say, of the order of one gram per cm.² Under a load 10,000 times greater, that is to say, 10 kilos per cm², the giving way of kaolin is produced with the same rapidity 400° lower.

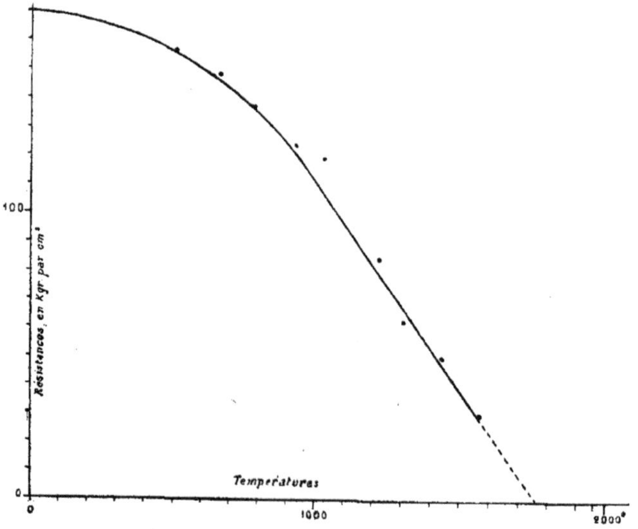

Fig. 1.

In order to explain the different way in which clay and quartz act. one could suppose that the latter has, on the contrary, a real point of fusion, without previous softening, and consequently independent of pressure. Our experiments have fully confirmed this prevision. When, towards 1,500°, a small cylinder of clay is crushed, it can be seen dilating into barrel shape, then completely flattening out into a thin cake with rounded edges without showing at any moment fracture properly so called.   When cooled the crushed mass preserved entirely its former hardness.   On the contrary, in the case of silica, the first operation of the pressure produces no appreciable effect, but, in increasing it gradually, the test piece breaks abruptly, showing the two sliding cones which are observed in all hard materials, when breaking under compression.   The broken fragments do not reunite at all during the cooling.   The effort necessary to produce this abrupt breakage decreases gradually with rise of temperature.   The table and the curve here annexed (Fig. 1) give a summary of the results of our experiments on a good silica brick of American make, marked " Star."   The resistances to crushing are expressed in kilogs. per cm^2—Table I.

TABLE I.

Temperature Degrees		Resistance to Crushing
15	..	170
520	..	158
670	..	150
800	..	139
950	..	125
1050	..	120
1200	..	85
1320	..	62
1460	..	50
1540	..	37
1600	..	30

These numbers lead by extrapolation to a resistance of 12 kil. at 1,700°, the usual temperature of the arches in steel furnaces. This resistance corresponds to about 10 times the effort sustained by the bricks in the arches.   Consequently, the stability of the latter is assured.   This mechanical resistance, being retained up to very high temperatures, is a special peculiarity of silica bricks.   It is not found, not only in clay bricks, but even in bricks whose essential constituent is magnesia, which is much less fusible still than silica.   Here is the reason of these differences. In both cases the bricks contain basic oxides foreign to the principal refractory material.   These oxides, alumina, lime, oxide of iron, alkalies, etc., produce in all cases a fusible matter already liquid towards 1,200°.   In the case of magnesia the solid grains float in the melted magma and slip easily one over another as wet sand would do.

Silica, on the contrary, at least in well-fired bricks, forms a continuous network in the pores of which the melted mass

lodges, like water lodges in the pores of pumice-stone without
diminishing the mechanical resistance.   The formation of this
network, in consequence of the recrystallization of the silica,
is due, as one of us has proved([1]) to the differences of solubility
of the different allotropic varieties of silica.   The quartz,
unstable at high temperature, dissolves in the melted magma
and recrystallizes first of all as cristobalite then as tridymite
(Fig. 2).

Fig. 2.—Showing a network of tridymite in a silica brick
very well fired.   Polarised light x 200.

This recrystallization of silica and subsequently the
formation of the permanent network, require firing sufficiently
prolonged at a suitable temperature.

Bricks little fired, whose network is not yet formed, are
composed of grains of quartz, floating in the melted mass;
they are plastic like the magnesia bricks or the clay bricks.
It is a fact well known in the manufactories that bricks little
fired are fusible and useless.

---

(1) H. Le Chatelier, *Revue de Métallurgie*, 13, 330, 1917.

When a good silica brick is heated, its resistance diminishes with rise of temperature. The reason of this is that the solubility of silica, increasing with the temperature, a progressive dissolution of the crystalline network is produced and tends to disintegrate it and even to dislocate it entirely. This effect is the more delayed as the network is better developed. It is on this that the quality of silica bricks depends before anything else. To study the factors on which the rigidity of this network depends is consequently a study of the first importance.

The researches of which we give an account to-day have for their object the solution of this problem. The principal factors to take into consideration are:

The proportion of fluxes.

The actual temperature of the brick. .

The good formation of the network.

The disintegration of the network through subsequent swelling out.

Proportion of fluxes. We have analysed some bricks which had been pointed out to us as having given particularly good practical results. The table below gives the weight per cent. of the basic oxides as well as the total weight of the sulphates obtained by an attack of hydrofluoric acid followed by evaporation to dryness after addition of sulphuric acid.

TABLE II.

Mark	Al_2O_3	Fe_2O_3	CaO	MgO & K_2O	Total	Sulphates
Star	0·94	0·15	1·79	0·50	3·38	9·06
Assailly	2·72	1·30	0 20	0·25	4·47	13·60
G. I.	1·02	0·60	1·48	0·00	3·10	8·40

The "Star" brick is the one whose resistance has been given as the highest. The "Assailly" brick, the manufacture of which goes back 30 years, had remained a year in the wall of a gas channel belonging to a steel furnace. It is on this sample that one of us([1]) recognised for the first time the trans-

---

[1] Le Chatelier, *Compt. Rend.*, 111, 123, 1890.

formation ot quartz into tridymite in bricks which have been heated for a long time. Finally, the brick " G.I. " is a brick of French manufacture, considered as one of the best amongst those actually manufactured by the French.

According to the above figures, good silica bricks contain between 3 and 5 per cent. of basic oxides, and the weight of sulphates varies from 8 to 14 per cent. The relation between the weight of the oxides and those of the sulphates varies necessarily according to the nature of the bases, but, as their relative proportions remain generally confined between fairly close limits, it can be said that the weights of the oxides represent a practically constant fraction of the weights of the sulphates, let us say, on an average of 35 per cent. This proportion of ingredients in sulphates can be made rapidly and suffices to estimate the composition of a silica brick.

Temperature. The temperature which the brick supports depends entirely on the use one wishes to make of it. In steel furnaces it must bear a temperature of 1,700°. Bricks offering the composition shown above possess, if they have been well manufactured, sufficient resistance to the temperature mentioned. Silica bricks are also used in ovens for the distillation of coal, in which the temperature is not so high. A proportion, at least double of the basic oxides, can be used without inconvenience for this purpose, and thus the manufacture is made much easier.

Constitution of the network. This is the most delicate part of the manufacture. In order to develop the network the brick must be kept for a very long time at such a high temperature that the melted magma may be sufficiently fluid. Experience seems to indicate, as the most favorable conditions, several days' firing at a temperature approaching 1,450°. This temperature must be inferior to that at which the quartz used is directly and rapidly transformed into cristobalite. The network is formed exclusively at the expense of the portions of silica which recrystallize through momentary dissolution into the melted mass. Finally this recrystallization will be accomplished the more completely and rapidly, everything being otherwise equal, if the quartz used in the fired mixture is fine or even very fine. However, a certain proportion of large grains is necessary to prevent the formation of cracks, whose propagation happens easily when the material is uniformly fine.

Disintegration of the network. When, after firing, grains of quartz remain not transformed, if the brick is afterwards

abruptly heated to a temperature at which quartz is rapidly
transformed, the expansion which accompanies this transformation shatters the network and robs the brick of all solidity.
Moreover. the pressure to which it is submitted in the arch
resists its lateral expansion and produces the phenomenon of
spalling; thus it falls in small fragments, causing sometimes,
in a few days, the destruction of an arch which should have
lasted months. The reproduced photograph (Fig. 3) shows a
similar brick before and after being placed in the furnace. The
cracks caused by this expansion are easily seen. The defect
is perhaps the gravest and certainly the most frequent of those
which affect bad silica bricks.

In the normal manufacture of a well-fired brick, the same
expansion is produced at the time of the direct transformation

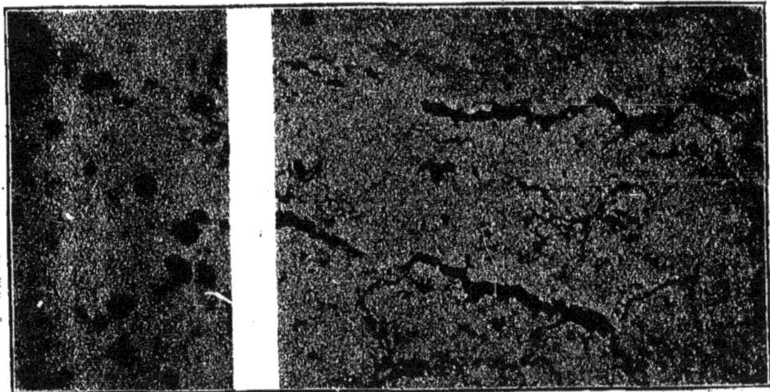

Fig. 3.—Showing insufficiently fired brick, spalling by rapid heating
in a steel furnace.

of the large grains of quartz, which never dissolve except on
the surface; but the inconvenience is not the same, because
then the expansion of the brick can develop itself freely in
all directions, and, on the other hand, the phenomenon is slow
enough to permit the network to reform itself on the places
where it has been cracked. Nevertheless, this unavoidable
phenomenon diminishes notably the mechanical resistance of
the brick.

Here are now the details of the experiments which have led
to the conclusions previously expressed. They refer to bricks
which have been handed to us by the manufactories of the

Navy, by the Ruelle foundry of guns and the Guérigny manufacture of armour-plating and for which they have given us the indication of qualities in use. We have measured the proportion of sulphates, the apparent and absolute densities, and the resistance to crushing in kilogs. per cm² measured at definite temperatures and after a definite time of firing. The results are shown in Table III.

TABLE III.

Marks	Quality	Sulphates Per Cent.	Absolute Density	Apparent Density	Resistance to Crushing		
					Temperature	Time	Crushing Strength
					Degrees		
Assailly	Very good	13·6	2·30	1·92	15		550
					1600	60'	90
Star ..	Very good	9·06	2·33	1 66	15		170
					1600	5'	33
					Id	60'	30
G.I. ..	Very good	8·40	2·33	1·88	15		185
					1600	60'	41
R.B. ..	Very good	13·1	2·35	1·60	15		62
					1600	60'	9·5
R.L. ..	Very good	14·3	2·40	1·85	15		265
					1600	5'	41
					Id	60'	25
G.A. ..	Good	14·0	2·40	1·77	15		190
					1600	60'	21
D. ..	Good	8·4	2·45	1·73	15		320
					1600	5'	55
					Id	60'	20
G.A.I.	Very bad	14·5	2·46	1·80	15		252
					1600	60'	4·4
R.F. ..	Very good	13·7	2·48	1·84	15		195
					1600		11
G.A. 2..	Very bad	12·8	2·48	1·78	15		148
					1600	60'	5
G.M. ..	Medium	9·5	2·53	1·84	15		84
					1600	5'	17
					Id	60'	2
R.L. ..	Bad (cracked)	9·75	2·56	1·94	15		350
					1600	5'	18
					Id	60'	4·5
R.S.G.	Bad (melts)	25·0	2·56	1·73	15		57
					1550	60'	22
					1600	melted	

B

These results allow us to formulate a very precise and practical conclusion. All good bricks possess at 1,600°, after heating for an hour, a resistance to crushing at least equal to 10 kil. per cm². The prolongation of heating, at this temperature, diminishes their resistance little in opposition to what happens with bad bricks.

DISCUSSION.

Dr. A. Scott (Glasgow):—When Fenner published his paper on the "Silica Minerals" some years ago, most people came to the conclusion that almost the last word had been spoken on the subject. According to Fenner α-quartz is the stable form up to 575°, β-quartz from 575° to 870°, tridymite from 870° to 1,470°, and cristobalite above 1,470°. The last named, however, may develop as an unstable intermediate phase below 1,470°. Recently, however, M. Le Chatelier has shown that the stability relations, particularly of tridymite and cristobalite, are still in doubt, as not only has he found tridymite in bricks which had been exposed for a long time at about 1,500° but he was also unable to convert tridymite to cristobalite by heating to 1,700°. Recently I have had occasion to examine a number of silica bricks which had been for many months at a temperature of over 1,500°, and my results confirm those of M. Le Chatelier. The hottest parts of the brick were entirely tridymite, but cristobalite mixed with tridymite appeared some distance from the hot end, while towards the cool end the number of partly converted unaltered quartz grains increased. This points to the conclusion that under certain conditions tridymite is stable above 1,500°.

This question is not merely an academic one but is of considerable technical importance on account of the volume changes which accompany the polymorphic transformations. Unfortunately the data we have regarding the physical properties of the various modifications are meagre and to a certain extent contradictory. Most text books on refractories, for example, give the specific gravities of the α-forms but are silent regarding the β or high temperature forms, which are perhaps the more important from the technical point of view. According to M. Le Chatelier, the transformation in the case of cristobalite is accompanied by a large volume change, while in the case of quartz and tridymite the volume change is much less. Fenner, on the other hand, states that the energy changes in the α-β transformations are very slight, which is scarcely consistent with a large volume change. What is really wanted is a revision of the determinations of the densities of the high temperature forms, but this is a problem involving considerable experimental difficulties.

The discrimination of cristobalite and tridymite is another point which may involve some difficulty, as, while in some cases it is possible to discriminate between the two by crystal habit, in others the only certain method is by means of the refractive indices. Although the refractive indices of the two minerals are near to each other, it is possible to discriminate between them by the immersion method, working under constant temperature conditions. The double refraction is of little use in distinguishing the two minerals. Recently quantitative estimations of the Rosiwal method of the various constituents present in silica bricks have been made, but it seems very doubtful if these possess any high degree of accuracy. The Rosiwal method is very useful in the case of igneous rocks which are comparatively coarse-grained, but where the grain size is small the measurements are of little value. In the case of silica bricks the grain size of the cristobalite and tridymite is small, these minerals are difficult to discriminate, and the presence of the bond introduces further complications, so that the Rosiwal method seems scarcely applicable.

With regard to unused bricks, it is often stated that they contain considerable quantities of tridymite or cristobalite— the two terms seem to be often regarded as interchangeable in this country—but, so far as my examinations have gone, neither form is present to any great extent in unused bricks. Under the microscope an unused brick shows that most of the quartz is still unaltered. It is only after repeated heating at a high temperature that the isolated quartz grains are converted, and even then the ganister fragments are only altered round their margins, while only after very prolonged use at a high temperature is anything like complete conversion obtained.

It seems to me that if anyone can find a method whereby the quartz can be converted to tridymite during the firing in the kiln many troubles will disappear, as the chief expansion will then occur before the brick is built into a furnace. This expansion, which occurs during the first few weeks of use, is of great importance, and in certain cases is greater than what theoretically would be expected from data we have concerning the densities of the various forms.

Professor FEARNSIDES (Sheffield):—In opening this discussion, Dr. Scott has made reference to a good many of the difficulties encountered by those of us who have endeavoured to interpret and apply to actual works conditions the information which has been published from the Geo-Physical Laboratory at Washington, respecting the phase-changes or transformations undergone by silica when heated, and, in following him, I should like to add my testimony to his appreciation of the immediate practical applicability of the results which M. Le Chatelier, whom we regard as the doyen of European phase-rule workers, has contributed to the available information upon this important subject.

With regard to the question whether tridymite or cristobalite represents the highest temperature phase, it is noteworthy that when molten steel is run into a mould built up of quartz-sand mixed with aluminous or ferruginous clay bond, the silica grains in the burnt skin of the mould, after the casting has cooled, are invariably found with a pellicle of cristobalite surrounding each cracked quartz-grain. I have looked for, but have not succeeded in finding, any development of tridymite in association with the sand which is so burnt by contact with steel castings, and which must have been raised instantaneously to temperatures between 1,500° and 1,600° C.

The idea introduced by M. Le Chatelier in this and in his earlier paper contributed to the *Revue de Métallurgie*, vol. 13, 1917, that tridymite is most conveniently formed from quartz by the intervention of some silicate flux or solvent, in which the quartz, at temperatures above the range of its true stability is dissolved, and out of which, when the solution becomes saturated with silica at that same temperature (which is within the stability range of tridymite), the tridymite can crystallise, is a very useful and acceptable working hypothesis. Almost a year ago, when describing "The Supplies of Refractory Materials available in the South Yorkshire Coalfield," and dealing more particularly with the siliceous seat-earth of the Halifax Hard Bed Coal of the Sheffield district, I ventured to attribute the quick-burning properties of true ganister to the presence of "some small quantity of impurity which behaves as a catalytic agent, and 'triggers off' the change from quartz into tridymite at temperatures comparatively low."[1] That use of the term "catalytic agent" has met with a good deal of adverse criticism, and I would like to

---

1 Trans. Inst. Min. Eng., vol. lii., p 270.

take this opportunity to acknowledge that my choice of term was not very happy, though as a matter of fact my conclusions and those arrived at by M. Le Chatelier are quite compatible.

I can confirm M. Le Chatelier's contention that the transformation change of the quartz particles into tridymite during the burning works invariably inwards from the outside of the grain, and with him I agree that it is of considerable importance that tridymite as a mineral crystallizes in the hexagonal (or pseudo-hexagonal) system, and generally forms in rods, which, in growing, tend to elongate themselves into needles, and so build up a skeletal network which is sufficiently rigid at high temperatures to support considerable stress. In this respect tridymite differs from cristobalite, which seems to belong to the cubic system, and when it crystallizes forms equi-dimensional grains, which, when associated with the melt, which is generally rather viscous, slide over one another with comparative ease.

My experience of the proportion of tridymite contained in silica bricks as generally sent out by the makers for use in steel furnaces is quite in accordance with what Dr. Scott has told us, and I have yet to find a brand of silica bricks in which the proportion of unaltered quartz averages less than 60 per cent. In many silica bricks, which have actually been in use in the arches of steel furnaces for a period which may have extended over many months, it is unusual to meet with specimens in which the whole of the quartz has been inverted, either into tridymite or into cristobalite.

Mr. G. W. MOTTRAM:—In nearly all research work on silica the starting point is either quartz, or quartzite, and this paper is no exception to the rule.

Very little is heard in any discussion on silica about the amorphous, or semi-amorphous forms, such as the various kinds of flint.

It would be interesting to learn what changes—if any—are effected in these forms of silica by heat treatment; whether there are changes to tridymite, or cristobalite, as in the case of quartz.

The calcination of flint, as pebbles, gives quite a different result as compared with the calcination of stratified flint.

While it is generally admitted that the burning of silica bricks should be carried to the inversion point so as to form tridymite, or cristobalite, and that only angular grains of silica should be used, one can hear of good results where these conditions have been absent in the manufacture. A short time ago I had a specimen silica brick, which had been made on the sand-lime brick principle, with $1\frac{3}{4}$ per cent. lime, presumably from sand with rounded grains, and hardened by steam. These bricks had not been burned, so we must assume that there was no change to tridymite, at any rate by ordinary heating. A number of these bricks were built in the wall of an open-hearth furnace, and the temperature gradually raised. The result was that they proved quite as durable and satisfactory as the standard quality silica bricks, made in the usual way, from Sheffield ganister. According to the usual theories these bricks should have changed to tridymite, or cristobalite, while in the furnace wall, and the resulting expansion should have shattered them.

It would almost seem as if the theory of the inversion points and the necessity for angular grains in the manufacture of silica bricks need further investigation.

Mr. J. G. ROBERTS (Barrhead):—There are one or two points which have been overlooked in this discussion:—

H. Le Chatelier makes one observation of very great importance, that is: That the density of cristobalite, tridymite, and vitreous silica are the same at $1,000°$ C. This is a point of great practical importance in the use of silica brick.

The greatest change in volume appears to take place during the transformation of quartz to cristobalite, and if this change has largely taken place then at high temperatures there should be little difficulty over undue expansion of the silica. We certainly want a good deal of the volume change brought about before the bricks are put into use.

Prof. Fearnsides referred to the transformation of quartz into cristobalite, and to a brick the portion of which nearest the heat zone consisted of cristobalite with tridymite further back. This does not agree with Le Chatelier's observation, who says that he took a brick which had been in a steel furnace and found it to consist entirely of tridymite. He tried to transform this into cristobalite but failed.

Le Chatelier appears to regard tridymite as the stable high temperature form of silica.

Mr. B. MOORE:—Calcined bone enters largely into the composition of English porcelain, attempts have been made to replace it with precipitated phosphates but they have been failures, the ware would not stand in the oven although the chemical composition was the same. The shape of the particles seems to have an important bearing on the practical results and confirm the views of the author of the paper.

Mr. J. HOLLAND (Sheffield):—I think Professor Le Chatelier's valuable paper is beyond criticism. We probably all agree on the main points, but on the question of the conversion of silica, if it is agreed that the ideal brick would be composed of tridymite, I do not think there is a single brickmaker who could not, if he tried, make a tridymite brick (probably at an excessive cost)—but could he get anyone to buy it? In silica bricks, as at present made, there is a certain amount of tridymite brought about in the ordinary course of manufacture, and every practical furnace builder, from experience, knows exactly how much to allow for expansion in building his furnace. There is no difficulty at all about this point. He does not worry about the conversion of the silica, but simply makes his expansion joints.

It seems to me to be a commercial question, whether we continue to make a silica brick at a comparatively low price, or are we to attempt to make tridymite bricks, for which probably no one would pay the price? Do we really need a tridymite brick? Is it necessary we should have them? If we take a silica brick which has been a long time in use in the furnace we find the end nearest the fire is cristobalite, and the tridymite follows after that, merging in the partly unconverted silica portion towards the outside end.

The present silica brick, in a properly regulated steel works, is so well understood by steel works managers, that they know exactly what to do with it. I do not know that if a furnace was built of tridymite brick it would last for an indefinite time, and as a brick manufacturer I have no desire to see a brick that will last for ever.

I am particularly interested in those drawings made by Professor Fearnsides—on the blackboard—arising out of one of the questions as to the difference in structure between ganister rock and silica sandstone. They typify the difference as we understand it in Sheffield.

The ganister stone, when ground—no matter to what degree of fineness—breaks with very irregular cleavage and sharp edges, and each particle is—what I call for want of a better term—prismatic in formation, so that when the heat is applied, and the conversion of the silica commences, these particles have a tendency to knit closely together, and thus form a mechanically strong brick.

There is a great difference in silica sandstone, which some people consider and describe as ganister. This stone, when broken and ground, is of a sandy character, and the small particles are round, so that, when conversion takes place, the particles have not that tendency to knit together that we find in ganister. I do not think that anything should be called ganister, unless it appromixates both chemically and physically to that mineral which has always been known as ganister.

In the first place, chemically: If the material is chemically bad there is an end of it; but if it be chemically good it does not follow that it would prove good in use unless it is also physically good.

Ganister, as we know it in Sheffield, is both chemically and physically good.

Fortunately, it is not too pure chemically, but contains a proportion of impurities which appear to be of great value in its application to industry. Its physical condition is ideal, and I repeat that only material approximating to these two conditions should be classed as ganister.

Mr. Arthur Acton, at our last meeting, made a suggestion that I think is one to which further consideration should be given, viz.: to classify material as:

    1. Ganister.

    2. Quartzite or Dinas type rocks.

And I would suggest a third: silica sandstone, and probably other types might be added. All these materials are of great industrial value, and, if the various rocks could be classified, and named as they most nearly approach a standard type, I think it would be a step in the right direction.

Dr. CECIL H. DESCH (Glasgow) said that the paper presented many points of great scientific interest, and it was gratifying that Prof. Le Chatelier, to whom we owed some of the earliest scientific investigations on this subject, had within the last few years again turned his attention to it, with results of great importance. It seemed desirable (perhaps the Ceramic Society had already considered the matter) that some general agreement should be reached as to the temperature which was to be regarded as characteristic of a refractory, in respect to its softening by heat. The Geo-Physical Laboratory, in accordance with physico-chemical theory, considered the melting point to be the temperature at which the crystalline structure was destroyed, and the substance passed into the amorphous condition. This was strictly correct, but had practical disadvantages. For instance, the melting point of silica was about

1,625° C., but the amorphous material formed at that temperature had a rigidity comparable with that of the solid, and one was met by the paradox that a steel furnace, built of bricks supposed to melt at 1,625° C., would be in regular use at temperatures considerably above that. A definition of the softening point, as distinguished from the melting point, would seem to be desirable.

The paper appeared to point to the advantage of firing at as high a temperature as possible. The fact that all the cementing materials used in making a silica brick had a softening effect suggested the possibility of eliminating them altogether. In the case of certain of the newer refractories he understood that it was possible to make a brick by binding particles of the highly-fired material with a paste of the same material in a plastic unfired condition. Could not this plan be applied to silica? Fragmentary silica, fired at so high a temperature as to have undergone inversion, might be bound by means of silica in a colloidal form. This was only a suggestion, which might perhaps be practicable.

Prof. Fearnsides' observation as to the effect of molten steel on sand in the foundry was not quite conclusive as to the relative stability of tridymite and cristobalite. The time of contact in such a case was short, and sufficient time was not allowed for equilibrium to be attained. Now, if a substance could exist in several polymorphic modifications, it did not necessarily follow that at any given temperature the new phase first formed would be that which was stable at that temperature. In fact, the contrary was frequently the case, and there was a rule, often found to hold good in physical chemistry, that the phase to be formed was not the most stable, but that next below it in stability, further inversion only taking place after a much longer time. This, of course, did not dispose of Prof. Fearnsides' argument, but made it uncertain whether any conclusion could be drawn, without extending the heating over a long period.

Mr. A. LINDSAY FORSTER (Glasgow):—One or two features connected with some silica bricks used by us may be of interest to the meeting. They were examined by Dr. Scott, who pointed out to me that the end of the brick which had been exposed to the furnace temperature was entirely tridymite, while the portion next to it was a mixture of tridymite and cristobalite, the outer cool end being almost entirely quartz. The brick might be considered as comprising three portions, that at the hot end being almost sharply distinguishable from the centre portion: it measured, say, half an inch, and was white and very "dry." These bricks had been in use for about 12 months at a furnace heat that was nothing extraordinary. They had "expanded" considerably, and the process of growth had continued for quite a long time.

If one knew how much "expansion" is to be allowed for with a particular kind or batch of bricks the user would be in a much more satisfactory position, but precise and reliable information on this point is not readily obtainable from makers as a rule.

Mr. WM. DONALD (Glasgow):—Mr. President, I would like to add some remarks to the subject as one who in the past manufactured silica bricks. With certain reservations, that I will make later, one might say that this is a branch of refractories that as a trade is being introduced to Scotland, at any rate its discussion is, and those present who have no knowledge of the manufacture of silica bricks must have been rather appalled, and those who have actually made them, looking at the subject from their practical point of view, must have been better able to understand the very learned discussions that have been put forward by Monsieur Le Chatelier, Dr. Desch, Professor Fearnsides and Dr. Scott. If they were to start with such highly abstruse scientific discussions to guide them I am afraid that silica brick manufacturers would feel they had a terrible load on their shoulders, and it is for that reason that I would like to add some practical suggestions.

In the forenoon we had a discussion about ganisters, and there was an attempt made to define what true ganisters are, one speaker suggesting that the name "ganister brick" should be limited to those bricks made in the neighbourhood of Sheffield from the Sheffield ganisters, and from other deposits clearly recognised by geologists to be from similar geological formations. I am certain that the Sheffield makers present will agree with me that from the various deposits available in the neighbourhood of Sheffield you could manufacture half a dozen qualities of silica bricks. You could make them of greater or less porosity, for instance. The trade is longest established in Sheffield, and it is for that reason I take it as standard. In the past it is true that the silica bricks made there were made from the finest deposits available. These naturally good deposits made bricks with certain qualities. The manufacture was begun on very simple lines at a time when the demand was not so urgent as it has been recently. Whatever plant was available was allowed to do its work in its own time so as to produce what in the course of years was found to be a good saleable article for furnace purposes. Generations followed each other in sixty years or so, and certain improvements had been devised throughout several periods of that time as the stress of passing circumstances quickened the minds of those taking part in the manufacture ; but it was hard for outsiders to notice much change. One good thing was that the quality remained on a sufficiently high level. The tremendous expansion of the trade during the past three years made it necessary for the miners and searchers to go further and further afield for new deposits of material to make the increased requirements of silica bricks of that quality. But the desired quality of the finished article was well known to all the workers employed at each stage of manufacture, and especially known to those who had the acceptance or refusal of the raw materials, and only those have been accepted that the existing plant could deal satisfactorily with.

The great expansion of the steel trade in Scotland, and the necessity to limit the railway traffic makes it desirable that an effort should be made to extend the trade more fully in Scotland. Already in more recent years there has been a well-established trade that produces a silica brick well known as suitable for special purposes in steel furnaces and coke ovens for instance, but which is easily recognised to be very different from those of Sheffield and South Wales. What is the difference? And why should other districts like those of the northern counties of England have their productions as clearly able to be recognised as having specific qualities and yet that makers should not be able to get results from their bricks exactly comparable to those obtained in the most trying parts of the furnaces by the Sheffield and Welsh makers? Why should the North Wales silica bricks differ from the South Wales bricks, and what are those differences? One very practical suggestion is to compare the physical characteristics such as weight and porosity of the best bricks able to be manufactured from the various deposits, making sure that the cubic contents is always the same. I remember doing this in the past before the war, and I found that bricks of exactly 100 cubic inches made by Sheffield makers weighed 56 cwts. per thousand at least; those by makers in the more northern counties weighed 52 cwts, per thousand, and those made at that time in Scotland from Scottish deposits of silica not more than 50 cwts. per thousand. In each case the analysis was practically the same, but there was a variation of porosity.

Let the definition of the raw silica material to be found in Scotland be ganister, silica stone or whatever its local name is, the quality of the deposits most desired—and we must ask our Geological Survey to find them—is that they must have high specific gravity without too much expansion, so that the bricks manufactured from them will have the substance of the Sheffield silica bricks.

In the manufacture of silica goods, where heat affects the dominant oxide and causes expansion with the increase of temperature, and contraction with falling temperatures, there must, I think, be a certain amount of porosity. It would be interesting to know from Dr. Desch if he has found that silica bricks completely transformed into tridymite or cristobalite remain unaltered with the variations of temperature that are experienced in open-hearth furnace parts. I have never seen such bricks, and until they have been proved to be more valuable than those that are allowed to undergo the transformation after their manufacture during steel furnace practice, I think the manufacturer may meantime take it that some porosity is essential to allow for the expansion. One factor that the users must consider, for instance, is that with expansion you can tighten the brickwork joints of a furnace in a way that is impossible under other circumstances, and that is a factor of great importance. With regard to this question of porosity we used to manufacture one quality of silica bricks which was so porous that it was of great value for fettling purposes, being able to be introduced with tongs direct into the furnace walls without disruption.

Used in the chequers these bricks could stand the constant heating and cooling, and after being six or eight months in use be taken out with long tongs, thrown on the ground, and to the extent of 50 or 60 per cent. be afterwards selected and rebuilt into the chequers again. These bricks, weighing 49 cwts. per thousand for size 100 cubic inches, had not the powers of sustaining the greatest loads and stresses in the most trying parts of the furnace that silica bricks have, that have greater substance and less porosity.

In the manufacture of silica bricks there must be purity, but, of greater importance still, there must be resistance to crushing strain under high temperatures, and, in the light of present day practice, there must, I think, be porosity.

Dr. J. W. MELLOR (communicated 10th Oct., 1917):— I have been greatly interested in the discussion which followed the reading of Monsieur le Chatelier's valuable paper. The equilibrium conditions of the many different forms of silica is an extremely important question and I do not suppose anyone realises this more than myself. I can also appreciate the position or attitude of the essentially practical man who will consider the report of this discussion to be possibly of academic interest, but may be bewildered with the array of the alpha and beta forms of quartz, tridymite, and cristobalite, as well as amorphous silica. He will want to know in plain language what it all means, so that if there is any useful practical application he can immediately get the subject in hand, because, after all has been said, it remains for the firebrick manufacturer to transform the theory into practical rules. I have been asked to translate our somewhat academic discussion on silica bricks into plain language. My explanation runs something like this: The protean forms of crystalline silica can be ranged in two classes: ordinary silica with a high, and well-fired silica with a low specific gravity. Weight for weight, the low specific gravity forms have the greater volume. We can call the high specific gravity forms A-quartz, and the low specific gravity form B-quartz. There are two important low specific gravity forms called tridymite and cristobalite, and the differences of opinion do not necessarily mean that one advocate is right and another wrong, but rather that the mechanism of the passage from the high to the low specific gravity form is not yet clearly understood. Mere assertion is not proof. It is clear that the high specific gravity silica slowly passes into the low specific gravity form at a high temperature. If the conversion is complete the bricks may expand up to about 12 per cent. We are satisfied that it is necessary for the firebrick manufacturer to burn his silica bricks at a high enough temperature, and for a long enough time to transform the greater proportion of the raw A-quartz into the less dense B-form. The manufacturer will, of course, say that this statement lacks novelty. At present the discussion is focussing attention on the importance of this subject, and those who like clear ideas want definite answers to the questions such as these: How far is it advisable for the firebrick manufacturer to attempt to carry out the transformation? Is there a general rule, or must each variety of silica be treated independently? The former question must be answered by the user, the latter by the manufacturer; but before these and other questions can be satisfactorily answered, it is for the academies to learn the exact mechanism of changes from one form of silica to the other. We therefore welcome academic discussions as a means to a definite end.

Mr C. N. FENNER (Washington, D.C.):—The paper "On the Refractory Properties of Silica," by MM. Le Chatelier and Bogitch, and the discussion by Professor Fearnsides and Dr. Scott, have reached me at a time when the demands by the Government for certain technical products required for military purposes have compelled me to lay aside research work and devote all my time and energies to meet these require- ments. This and the fact that I am so situated at present that no references are available compel me to limit my contribution to the discussion to a few lines only.

Dr. Scott has kindly referred to my work on the silica minerals,[1] but his remarks indicate that some of the conditions which affect the attainment of stability, on which I laid emphasis, were not fully in his mind at the time he wrote. The fact mentioned by him that tridymite is found in bricks which have been exposed to a temperature of over 1,500° is, I believe, not at all contradictory of my conclusions, but is quite consistent with the extraordinary sluggishness of inversion which I found to be manifested as a general feature of the silica minerals and to which I directed especial attention. In such bricks as he speaks of it is very generally the case that we find not only tridymite but also quartz associated with the cristobalite which is present. Of course no one will assert that all three minerals are stable at the same time, and reflection should indicate to us that the conditions for determining stability relations are not of the best when we expose silica bricks (of somewhat variable composition, especially as regards the per- centage of fluxing binder) to the fluctuating temperature con- ditions found in furnaces in manufacturing plants.

Several investigators, working on silica refractories, have confirmed my conclusions, and before serious question of their correctness is raised I think that evidence should be brought out that either the principles on which I based these con- clusions do not apply or that the results which I described cannot be obtained by others. Of course in any experimental work of this kind all the conditions should be carefully controlled.

Professor Fearnsides makes mention in his discussion of an idea which he attributes to M. Le Chatelier "that tridymite is most conveniently formed from quartz by the intervention of some silicate flux or solvent, in which the quartz, at temper- atures above the range of its true stability is dissolved, and out of which, when the solution becomes saturated with silica at that same temperature (which is within the stability range of tridymite) the tridymite can crystallize." Anyone interested may find this matter gone into rather fully in my paper.

I can hardly agree with Dr. Scott's opinion that there should be a parallelism between the magnitude of the energy-change and that of the volume-change involved in transformation.

---

[1] The Stability Relations of the Silica Minerals, *Amer. Journ. Sci.*, xxxvi., 331-384, Oct.. 1913.

M. H. Le CHATELIER :—I am very pleased my paper has provoked such thorough discussion. On nearly all points I am in agreement with the previous speakers.

Perhaps it may not be necessary to pursue fresh researches on the densities of the different varieties of silica at high temperatures, as Dr. Scott desires. They can indeed be calculated very accurately from the densities measured at ordinary temperature, using the coefficients of expansion. In this way it is found that cristobalite possesses a density towards $1,000^\circ$ bordering on $2\cdot22$. The difficulty of determining quantitatively the proportions of tridymite and cristobalite, and still more the impossibility of obtaining complete transformation of the quartz into tridymite in the first firing of the bricks, is acknowledged.

This transformation requires weeks to finish it in a steel furnace, where the temperature is much more elevated.

Prof. Fearnsides points out quite correctly a fact in the subject of the first transformation of quartz into cristobalite under the action of heat. After heatings of short duration, followed by rapid cooling, cristobalite only is always observed. Thus fine crystals of cristobalite are found in the wall of a Bessemer converter, but little or no tridymite, because the firings are so little prolonged and the coolings rapid. These conditions are all different from those of the Siemens furnace, where the temperature is maintained very high almost constantly. It is only then that tridymite is regularly developed

As to Mr. Mottram's question concerning flint, this variety of silica is transformed, like quartz, first into cristobalite, but it is effected at a very low temperature, from $1,150^\circ$, well before fusion of the silicates begins. In the absence of binding material, the swelling becomes enormous and the very porous brick loses all resistance. It does not seem that good silica bricks have bene made successfully hitherto by using flint exclusively.

It is obviously not impossible to employ directly sand-lime bricks with small content in lime, on condition of conducting slowly the heating of the furnace, in such a manner as to fire the brick in its place. But it is questionable whether this practice is advantageous, because if the parts most heated are well fired, the intervening portions of the brick, which support temperatures between 500° and $1,200^\circ$, lose all stability and are then liable to disintegrate under the operation of the inevitable movements due to expansion of the heated walls.

The question of knowing if the angular grains of a ground quartz rock can be replaced by round grains of sand requires fresh study. Trials made with a mixture of 75 per cent. of

Fontainebleau sand and 25 per cent. impalpable silica, passed by a tube mill, have given encouraging results.

Mr. J. Holland protests with good reason against the notion of seeking to make bricks of silica transformed completely into tridymite. The author of the paper has never countenanced such an opinion. It would be absolutely unrealizable, since for this transformation a stay of one month in a steel furnace at a temperature above 1,600° would be necessary, although the most powerful industrial firings of silica bricks reaches only 1,400, and firing at the highest temperature often last only a few hours, never more than a few days. Yet it seems necessary to obtain a sufficient proportion of tridymite, that which can be reached by the introduction into the paste of a sufficient proportion of impalpable quartz, that is to say, running through a sieve of 4,900 meshes to the square cm.

The information given by Dr. Cecil Desch on the subject of the fusibility of silica will require to be discussed very closely. The different varieties of silica (quartz, cristobalite, and tridymite) necessarily have different fusion points, tridymite melting at the highest temperature. It seems difficult to admit that its fusion point can be brought down to 1,625°. It cannot be observed even at 1,700°. In old bricks completely transformed into tridymite, and heated to that temperature, a part of the mass is seen to become vitreous by solution of a certain quantity of tridymite in the metallic silicates; but the parts respected by the flux remain absolutely unaltered as to structure and dimensions of the crystals. Sometimes 1,480° is cited as the fusion point of quartz. That number appears no more established on exact bases. It is probable that the fusion point of tridymite—or, if preferred, its passage to the amorphous condition—is bordering on 1,800°.

Mr. A. Lindsay Forster has described with great accuracy the successive appearances of a silica brick examined in regions more and more distant from the heated portion. The arch of a steel furnace is generally impregnated, on the lower part, with ferruginous slags, and it is in this region that the crystals of tridymite are most completely developed.

Mr. W. Donald is terrified at the burden of scientific data under which it is intended to crush the manufacturers of silica bricks. Without in the least contesting the results to which these manufacturers have attained by a long experience, extending over several generations, it will be permitted to remark that, in the absence of that practical experience, laboratory investigations, by their scientific studies, attain to the production after a few months of work, of as good bricks as those of which the present quality, by purely empirical methods, required long years of groping. That is the essential object of scientific methods of working. As for the differences pointed out between Scotch and Sheffield bricks, doubtless it would be easy to find out the cause quickly if samples were at hand to be compared. It is by no means proved that rocks of a special nature must be employed in order to have good bricks. Studies made hitherto show, on the contrary, that with every quartz of sufficient hardness and not reducing to powder in the fire, it is possible and even easy to make good bricks on condition of employing a firing process suited to each kind of stone.

The explanation of Dr. Mellor shows very clearly what the object of scientific studies is, and how they can be interpreted in order to render them accessible to manufacturers. Perhaps, however, he has not laid sufficient stress on one peculiarity of great importance. One of the varieties of silica, of small density, presents a very curious property, but also very injurious. It experiences about $225°$ an abrupt change of dimensions which brings about fractures when the bricks are heated too rapidly. It cannot be recommended too strongly to metallurgists to set the fire very slowly in furnaces constructed of silica bricks. In default of this precaution, furnaces constructed of bricks, excellent from every other point of view, are exposed to being ruined from the first moments of heating.

With regard to Mr. Fenner's assertion that tridymite ceases to be stable at the highest temperatures of the furnaces, it may be said that the daily practice of steel furnaces contradicted it long ago. The most strongly heated parts of the bricks, originally composed of a mixture of quartz and cristobalite with very little tridymite, is progressively transformed and in a complete manner into this last variety of silica. Moreover, the experiment is very simple to realize in the laboratory. It is sufficient to take a brick almost completely in the condition of cristobalite and to heat it for several hours at $1,700°$, in order to promote an abundant development of tridymite, and that without any of the alternations of heating and cooling to which Mr. Fenner makes allusion. His personal experiments have never given proof to the contrary. They are certainly very exact, but have been badly interpreted by their author. On heating tridymite at a high temperature, in the presence of a flux, and cooling the mass slowly, cristobalite is certainly found, even well crystallized cristobalite, as in numerous examples I have given. But this tridymite is not produced by heat; it has crystallized during cooling, separating from the melted magma in which an excess of silica is dissolved, thanks to the elevation of the temperature. Every time that

C

a glass highly charged with silica is cooled with moderate speed, the first variety which crystallizes is cristobalite. So, if red iodide of mercury be heated in boiling alcohol, that is to say, well below the point of transformation of the iodide into its yellow variety, and be cooled after solution, the yellow variety is always produced. It is not concluded that the point of transformation of the red iodide into the yellow iodide is below the boiling point of alcohol. It really corresponds to the temperature of 125°. It is exactly the same with cristobalite. In order to render Mr. Fenner's experiment conclusive, the sample should have been steeped, after heating, in order to prevent all new crystallization of silica on cooling.

XV. — Sur les propriétés réfractaires de la Magnésie.

MM. H. LE CHATELIER et B. BOGITCH.

LA magnésie, dont les propriétés réfractaires ont depuis longtemps été signalées par M. Schloesing, est aujourd'hui couramment employée à la fabrication de matériaux réfractaires très réputés. Leur emploi dans les fours d'aciérie s'est rapidement généralisé parallèlement au développement des procèdés basiques. Dans ce mode de traitement, l'affinage du métal est obtenu en présence d'un laitier riche en chaux, c'est à dire très basique, d'où le nom du procédé. Grâce à cette teneur élevée en chaux, on peut éliminer de la fonte, non seulement le carbone, le silicium et le manganèse, mais encore le phosphore qu'il serait impossible de faire disparaitre en présence d'une scorie riche en silice.

L'emploi des laitiers basiques serait impossible dans un four dont les parois seraient entièrement construites en briques siliceuses on argileuses ; ces matériaux se dissoudraient trop rapidement dans le bain calcaire. Les briques de magnésie, au contraire, constituées elles mêmes par un oxyde basique, résistent parfaitement dans ces conditions. Dans tous les fours basiques, la partie inférieure des parois verticales et souvent la sole elle même sont construites en magnésie. La voute, par contre, est toujours faite avec des briques de silice.

Les briques de magnésie passent pour être très refractaires ; la magnésie pure fond seulement vers 2,400° c'est à dire à une température supérieure de 700° à celle des fours d'aciérie ; mais la magnésie employée pour la fabrication des briques n'est jamais pure. Elle renferme des proportions variables d'oxyde de fer qui colorent plus ou moins fortement les briques en brun. Ce fer se trouve dans le minerais à l'ètat de carbonate de fer isomorphiquement mèlé au carbonate de magnésie naturel. Elle contient encore de la silice et un peu d'alumine provenant soit de silicates magnésiens associés au carbonate, soit des cendres du combustible employé dans la

première cuisson. Toutes ces impuretés augmentent nècess-
airement la fusibilité de la masse.

Il nous a paru intéressant d'étudier les propriétés réfractaires des briques de magnésie comme nous l'avions fait
précédemment pour celles d'argile et de silice. La méthode
expérimentale employée a été la même que celle de nos
premières recherches.

Nous donnerons par la même occasion des résultats
relatifs à une brique de fer chrômé. On emploie ces briques
dans la construction des fours pour séparer les briques de
magnésie de celles de silice. Le contact direct de matériaux
basiques et acides provoquerait leur fusion mutuelle. Le fer
chrômé au contraire ne reagit ni sur la silice, ni sur la magnésie.

Les matières essayées ont été, soit les briques industrielles, soit des échantillons prèparés au laboratoire. Pour
obtenir un terme de comparaison, on avait essayé de préparer
un bloc de magnésie très pure en fondant au four électrique
de la magnésie precipitée ; mais à la haute température
nècessaire pour cette fusion, la chaux des parois du four et
les impuretés qu'elle renferme se volatilisent et vont souiller
le produit fondu. On a obtenu ainsi une matière renfermant
seulement 94 centièmes de magnésie et tout juste comparable
aux bonnes briques de fabrication industrielle.

Nous donnerons dans un premier tableau la liste des
produits essayés, dans un second, leur analyse chimique et
dans un troisième, leur résistance à l'ècrasement mesuré à
différentes températures.

Liste des produits essayés.

I.   Brique de Styrie, cuite à 1,450°, fabriquée en 1890,
qualité normale.

II.   Bonne brique d'Eubée, fabriquée en 1910.

III.   Brique *G*, bonne fabrication actuelle.

IV.   Brique *B*, mèdiocre, addition dans le pâte de
3 centièmes de pyrite grillée.

V.   Magnésie pure fondue au four èlectrique et souillée
par cette opération.

VI.   Matière première des briques *B*, agglomérée au
foui èlectrique.

VII.   Brique de fer chrômé.

ANALYSE CHIMIQUE.

	I	II	III.	IV	V	VI	VII
Magnésie	86·7	93·4	89·4	81·2	93·7	88·5	12·3
Chaux	1·0	3·7	4·5	4·8	2·7	4·5	5·8
Oxyde de fer (Fe_2O_3)	6·0	0·5	1·1	4·2	0·3	1·4	15·5 (FeO)
Alumine	0·6	0·2	0·8	1·0	1·1	0·0	10·9
Silice	6·7	2·8	4·2	8·8	3·2	6·5	4·7
Oxyde de chrôme ..	—	—	—	—	—	—	50·0
„ de Manganèse	—	—	—	—	—	—	1·5
Total ...	101·0	100·6	100·0	100·0	101·0	100·9	100·7

RÉSISTANCE A L'ÉCRASEMENT EN KG. PAR CM².

Température	15°	1000°	1300°	1500°	1600°
I	145	85	66	3·6	1·8
II	420	—	—	185	8
III	390	—	—	> 90	4·8
IV	230	—	—	16	3·5
V	—	—	—	> 90	6·6
VI	530	—	—	—	3·5
VII	260	120	6	2	1

Pour les deux premières briques, celles de Styrie et d'Eubée, les expériences ont été en réalité plus nombréuses que celles portées au tableau ci-dessus. Elles ont permis de tracer la courbe complete des résistances mècaniques (Fig. 1), les deux courbes sont caractérisées par une chute brusque de résistance à l'écrasement qui se produit entre 1,300° et 1,400° pour la brique de Styrie, la moins pure des deux, entre 1,500° et 1,600° pour celle d'Eubée. Toutes les briques de magnésie présentent cette chute brusque de résistance à une température plus ou moins élevée suivant leur degré de pureté. Tout se passe comme si, à une certaine température, les matières étrangères fondaient brusquement de façon à laisser les grains de magnésie isolés dans un magma fondu. Ils sont alors dans le même état que du sable humide et ne possèdent plus qu'une résistance mécanique très faible. Les meilleures briques de magnésie présentent à 1,600° une résistance à l'ècrasement bien inférieure à celle des bonnes briques de silice. De plus, à ces températures élevées la déformation des briques de magnésie se fait comme pour celles d'argile, elles cèdent progressivement au lieu de se rompre brusquement comme les briques de silice. Pendant le réfroidissement la matière ècrasée se ressoude et reprend sa dureté après la solidification du magma fondu.

Ces résultats expliquent comment les briques de magnésie résistent moins bien dans les parois des fours que celles de silice, bien que leur température de fusion, lorsqu'on l'observe en dehors de tout effort mècanique, soit très notablement supérieure, 2,050° au lieu de 1,750°.

L'allure de la chute de résistance dans la brique de fer chrômé est analogue à celle des briques de magnésie, mais avec une température beaucoup plus basse pour la perte rapide de solidité, 1,100° au lieu de 1,350° à 1,550° suivant la pureté de la magnésie.

[TRANSLATION BY J. A. A.]

MAGNESIA, the refractory propèrties of which were long ago pointed out by Schloesing, is to-day readily employed in the manufacture of refractory materials of high repute. Their employment in steel furnaces has rapidly become general, parallel with the development of basic processes. In this mode of treatment, the refining of the metal is obtained in the presence of a slag rich in lime, that is to say, very basic, whence the name of the process. Thanks to this high content in lime, there can be eliminated in the smelting not only carbon, silicon, and manganese, but moreover phosphorus, which it would be impossible to cause to disappear in the presence of a slag rich in silica.

The employment of basic slags would be impossible in a furnace the walls of which were constructed entirely of silica bricks or clay bricks, these materials dissolving very rapidly in the calcareous bath. Magnesia bricks, on the contrary, themselves constituted of a basic oxide, resist perfectly in these circumstances. In all basic furnaces the lower part of the vertical walls, and often the sole itself, are constructed of magnesia. The arch, as a set-off, is always made with silica bricks.

Magnesia bricks are considered to be very refractory; pure magnesia melts only about 2,400°, that is to say, at a temperature 700° above that of steel furnaces; but magnesia employed for the manufacture of bricks is not always pure.

It includes variable proportions of iron oxide, which colours the bricks more or less strongly brown. The iron is found in the mine in the state of carbonate of iron, isomorphously mixed with the natural carbonate of magnesia. It contains moreover silica and a little alumina arising either from magnesian silicates associated with the carbonate, or from ashes of the fuel employed in the first burning. All these impurities necessarily increase the fusibility of the mass.

It appeared interesting to us to study the refractory properties of magnesia bricks like we had done previously for clay bricks and silica bricks. The experimental method employed has been the same as that of our former researches.

We will give on the same opportunity results relative to a chromite brick. These bricks are employed in the construction of furnaces for separating magnesia bricks from silica bricks. Direct contact of basic and acid materials would promote their mutual fusion. Chromite, on the contrary, reacts neither with silica nor magnesia. The substances tried have been either industrial bricks, or specimens prepared in the laboratory. In order to obtain a term for comparison we tried to prepare a block of very pure magnesia by melting precipitated magnesia in an electric furnace; but at the high temperature necessary for this fusion the lime of the furnace walls and the impurities which it included volatilise and go to contaminate the melted product. We thus obtained a substance including only 94 per cent. of magnesia and exactly comparable to good bricks of industrial manufacture.

In a first table we will give the list of products investigated, in a second their chemical analysis, and in a third their resistance to crushing, measured at different temperatures.

List of products investigated.

I. Styrian brick, burned at 1,450°, made in 1890, quality normal.

II. Good Eubœan brick (Greek), made in 1910.

III. Brick *G*, good present-day manufacture.

IV. Brick *B*, ordinary, addition in the paste of 3 per cent. roasted pyrites.

V. Pure magnesia melted in an electric furnace, and contaminated by that operation.

VI. Principal substance of bricks *B*, agglomerated in the electric furnace.

VII. Chromite brick.

CHEMICAL ANALYSIS.

	I	II	III	IV	V	VI	VII
Magnesia	86·7	93·4	89·4	81·2	93·7	88·5	12·3
Lime	1·0	3·7	4·5	4·8	2·7	4·5	5·8
Iron Oxide (Fe_2O_3)	6·0	0·5	1·1	·2	0·3	1·4	15·5 (FeO)
Alumina	0·6	0·2	0·8	1·0	1·1	0·0	10·9
Silica	6·7	2·8	4·2	8·8	3·2	6·5	4·7
Chromium oxide ...	—	—	—	—	—	—	50·0
Manganese oxide ...	—	—	—	—	—	—	1·5
Total ...	101·0	100·6	100·0	100·0	101·0	100·9	100·7

RESISTANCE TO CRUSHING (KILOG. PER SQ. CM.).

Temperature	15°	1000°	1300°	1500°	1600°
I	145	85	66	3·6	1·8
II	420	—	—	185	8
III	390	—	—	> 90	4·8
IV	230	—	—	16	3·5
V	—	—	—	> 90	6·6
VI	530	—	—	—	3·5
VII	260	120	6	2	1

For the first two bricks (Styrian and Eubœan) the trials
have been in reality more numerous than those contained in
the above table. They have allowed the complete curve of
the mechanical resistances (Fig. 1) to be traced. These two
curves are characterized by an abrupt fall in resistance to
crushing, which is produced between 1,300° and 1,400° for
the Styrian brick, the less pure of the two, between 1,500°
and 1,600° for the Eubœan brick. All the magnesia bricks
present that abrupt fall in resistance at a temperature more
or less elevated according to their degree of purity. Everything happens as if at a certain temperature, the foreign substances melt suddenly so as to leave the magnesia grains
isolated in a melted magma. They are then in the same
condition as wet sand and possess no more than a very weak
mechanical resistance. The best magnesia bricks present at
1,600° a resistance under load much inferior to that of good
silica bricks. Moreover, at these high temperatures deformation
of magnesia bricks takes place as for clay bricks, they yield
progressively instead of breaking suddenly like silica bricks.

ᴌuring the cooling the crushed material becomes cemented together again and resumes its hardness after the solidification of the melted magma.

These results explain why magnesia bricks offer less resistance in the walls of furnaces than silica bricks, although their temperature of fusion, when it is observed outside of all

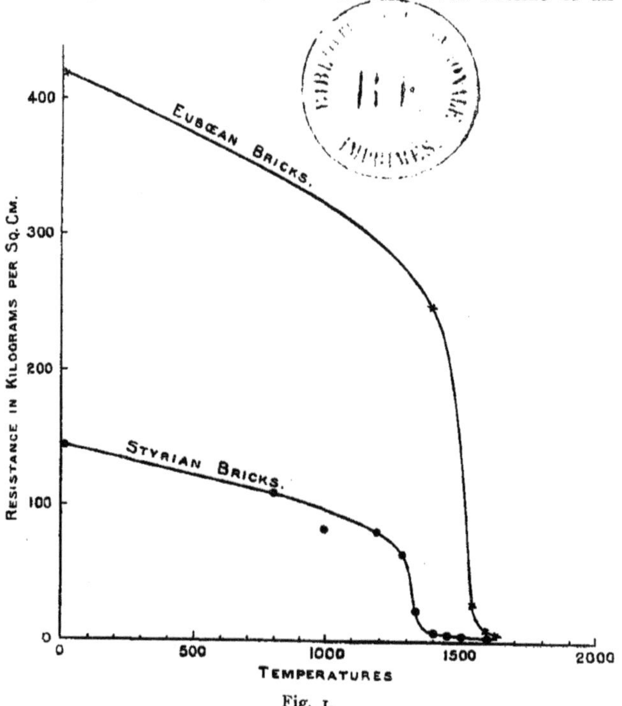

Fig. 1.

mechanical effort, is very notably superior, 2,050° instead of 1,750°

The velocity of the fall in resistance in a chromite brick is analogous to that of magnesia bricks; but with a temperature much lower for the rapid loss of stability, 1,100° instead of 1,350° to 1,550°, according to the purity of the magnesia.

www.ingramcontent.com/pod-product-compliance
Ingram Content Group UK Ltd.
Pitfield, Milton Keynes, MK11 3LW, UK
UKHW022218070726
13613UKWH00004B/1734